ISBN 978-0-265-84186-0
PIBN 10897625

INVESTIGATIONS IN AVOCADO BREEDING

by

B. S. Nirody, B.A.,F.R.H.S.

Thesis submitted for the Degree of Master of Science
Massachusetts Agricultural College

Amherst, Massachusetts
April, 1922

OUTLINE

1. Introductory

2. Botanical description of the avocado (Persea americana Mill.) and its three horticultural groups or "races".

3. Merits and demerits of each type.

4. The essential horticultural qualities of a standard avocado.

5. Analysis of the qualities of the different varieties now in cultivation.

6. Qualities which are sought to be counterbalanced.

7. Some combinations for cross-pollination with reasons for such.

8. Details of pollination.

9. Tabular statement showing the time of day at which the flowers of each variety present optimum conditions for pollination.

10. The nature of the result.

Supplementary

11. Varieties which should be interplanted with others in order to increase the setting of fruit and obtain a better yield.

12. Summary

13. Appendix.

14. Bibliography

15. Illustrations

INVESTIGATIONS IN AVOCADO BREEDING

1. **Introductory**

It is now generally known that the culture of the
avocado (Persea americana Mill.) already bids fair to rival
the citrus industry in Florida, and that a similar outlook
is described by the growers of California. The annual reports
of the Florida State Horticultural Society and of the California Avocado Association in successive years make it abun-
dantly clear that the avocado has already established its
position as a standard fruit of commercial importance.
Mr. Wilson Popenoe, who has made an extensive study of the
fruit in all those countries where it has been grown since
early times - the West Indies, Mexico, Guatemala, Costa Rica
and northern South America - and who is our best authority on
this subject, is convinced that the avocado will before long
be as familiar to the American people as the banana is today.[*]

While the incredible prices which the fruit commands
today, because of the inadequate supply, make it one of the most
profitable branches of subtropical orcharding, growers both in
California and Florida agree that from a strictly commercial
point of view, it is necessary to stabilize the industry by
establishing standard varieties of the fruit which best meet

---

[*]For an analysis of the fruit indicating its high food
value see Appendix.

the requirements of the growers as well as of the market.
This is voiced most emphatically by the owners of avocado
orchards at their annual conventions year after year. In
other words, the avocado industry is now in the same stage
of development as the citrus industry of a few years ago.
In citrus fruit culture commercial planting is confined to
one or two standard varieties of the orange, the grapefruit
and the lemon. Likewise, in avocado orcharding from among
the countless varieties that are introduced from regions where
the fruit is native, the best types that answer the needs of
the different climatic conditions of Florida or of California,
and the requirements of the northern markets will have to be
evolved by breeding. Once this ideal type is produced, all
subsequent propagation can be carried on by asexual methods
and the desirable qualities can be perpetuated.

The cultivated avocados fall naturally into three
distinct groups or "races": (1) The West Indian, (2) the
Guatemalan and (3) the Mexican. Each of these has important
merits as well as defects; these are discussed in detail in
another part of this paper. Briefly, the West Indian thrives
well in southern Florida, but not in northern Florida, or in
California. The Mexican, on the other hand, is known to grow
well in California and certain sections of northern Florida,
but is not well adapted to conditions in southern Florida.

The Guatemalan has qualities which make it of great value
both in Florida and in California. But even the best exist-
ing varieties of the Guatemalan group have defects which the
grower would like to eliminate - hence the necessity of breeding, so as to combine the qualities of the best Guatemalan
and Mexican varieties for the California orchards and the best
Guatemalan and West Indians for those of Florida.

The present investigation was undertaken under
Florida conditions. For the facilities provided by an extensive
orchard containing several hundred trees of the leading varieties
in bearing condition and also for valuable information regarding
the performance of those trees in previous years, the writer
is indebted to Mr. William J. Krome of Homestead, Florida,
Vice-President of the Florida State Horticultural Society.

2. Botanical Description

"The genus, Persea, to which the avocado belongs is
a member of the family, Lauraceae. Among the other more
important economic members of the family are cinnamon (Cinnamomum
cinnamomum (L.) Cockerell), camphor (Cinnamomum camphora (L.)
Ness.), and sassafras (Sassafras sassafras (L.) Karst.) With
the exception of cinnamon, they are chiefly used in medicine.
The avocado is the only member of the family cultivated for
its edible fruit.

Mez, in his monograph of the family (Lauraceae
americanae, 1889) describes forty-seven species of Persea and
states that the genus is confined to the American continent,

with the exception of one species in the Canary Islands.

On the contrary, F. Pax (Engler and Prantl, Die Natürlichen Pflanzenfamilien, 1889, 3:2 : 114-115) restricts the genus to ten species, only one of which, P. persea (L.) Cockerell (P. gratissima of Pax), belongs in America" (Bull. 77, Bur. Pl. Ind.)

It is outside the purpose of this investigation to enter into the controversy as to whether the three so-called "races" of the avocado, (the West Indian, the Guatemalan,and the Mexican), belong to one species, Persea gratissima, Gaertn f. (P. americana, Mill.) or whether the West Indian and the Guatemalan alone comprise the Persea americana, and the Mexican group forms a separate and distinct species, which, according to Mez, is Persea drymifolia (Cham. & Schlect). Let us suppose for the time being that all three are one species. The different groups may be distinguished as follows:

"A.   Leaves - Anise-scented; skin of fruit thin
                 and soft .... Mexican

AA.   Leaves - Not anise-scented; skin of fruit thick.

B.   Surface of fruit usually smooth; skin leathery,
                 usually not more than 1/16 in. thick;
                 seed coats frequently distinct, the
                 outer one adhering to wall of seed-
                 cavity; cotyledons often rough.:....
                                                West Indian

XB. Surface of fruit usually rough or warty;
skin brittle, granular 1/16 - 3/16 in.
thick; seed coats adhering closely to
the nearly smooth cotyledons ....
Guatemalan"

Bailey, L. H., Cyc. of Hort. 1916, Vol. 5,

P. 2556.

3. From a horticultural standpoint each of the three groups mentioned
above has merits as well as demerits and these may be summarized
as follows:

The West Indian Group

Merits		Defects
1. Heat resistant.	:	1. Susceptible to frost.
Thrives well in Florida.	:	2. Shipping qualities comparatively
2. Named varieties are of	:	poor.
good quality.	:	3. Seed often loose in the seed
3. Prolific.	:	cavity.
	:	4. Matures too early in the season.
	:	The last mentioned defect is con-
	:	sidered by far the most important
	:	one commercially.

The Guatemalan Group

Merits	Defects
1. Matures as a rule in a later season.	1. Susceptible to heat. Tips of leaves are often scorched by heat in Florida.
2. Resistant to frost.	
3. Includes varieties of very good quality and of vigorous growth.	2. Irregular bearer, often bearing in alternate years.
4. Seed tight in the seed cavity.	3. Skin usually rough and often warty in appearance.

. . . .

The Mexican Group

Merits	Defects
1. Very productive.	1. Fruit usually small with large seed.
2. Hardy.	
3. Rich in taste with a distinct flavor.	2. Shipping qualities poor.
	3. Not well adapted to the lowlands.

4.   It is here pertinent to inquire as to the qualities most
desired in a standard avocado from a horticultural standpoint;
They are these:

    1. The tree must be resistant to frost injury.

    2. The tree must be a vigorous grower.

    3. It must be a good bearer, i. e., yielding reasonably
heavy crops.

    4. It must bear regularly, i. e., annually.

    5. It must bear when fairly young, i. e., it must not take
too long to come into bearing.

    6. The fruit must mature in the right season, i. e.,
from October to the middle of March. This is by far
the most important consideration with the growers,
since fruit marketed outside this period brings poor
prices because of competition with imported fruit in
the northern markets.

    7. The fruit should be of medium size.

    8. The fruit should be uniform in shape, preferably round,
instead of elongate or pyriform, to permit of standard
packages with a definite number to the crate.

    9. It must be disease-resistant. Some varieties are
susceptible to scab.

   10. The quality of the fruit should be good; in practice
this offers no material difficulty, since the varieties
already in cultivation are nearly all of good quality
and free from fibre.

11. The seed should be tight in the cavity; a loose
    seed injures the fruit in transit.

12. The size of the seed should be small compared with
    the size of the fruit. This does not present any
    difficulty in practice, since in a medium, standard
    sized fruit, the size of the seed can usually be
    overlooked unless it is abnormally large.

13. The ripening of the fruit should be even. Some
    fruit of the Mexican group cultivated in Florida
    is often known to ripen unevenly.

14. The skin should be smooth but hard enough to ship
    well. Some Guatemalan avocados have a rough, warty
    and ungainly appearance, while the smooth skinned
    West Indian and Mexican fruits do not ship so well.

15. The color of the fruit under present market require-
    ments should preferably be green. However, some
    of the purple Guatemalan varieties are generally
    associated with excellent quality and flavor. This
    question of color had perhaps be best overlooked
    at present, because it is not unlikely that at a
    future date, when public opinion is better informed
    about the quality of named varieties, the purple
    colored fruit may even be preferred to the green.

It will be seen that the choicest of the West Indian,
Guatemalan or Mexican avocados now in the trade, while conforming
to the ideal in many important particulars, nevertheless have an

outstanding defect or two which the grower feels ought to be eliminated. These defects are more or less characteristic of the group or "race" to which they belong.

The following is a list of the varieties considered for breeding in the present investigation:

W. I. = West Indian   G = Guatemalan   M = Mexican

1.	Collins (G)	7.	Queen (G)
2.	Fuerte (Hybrid: G & M)	8.	Sharpless (G)
3.	Linda (G)	9.	Taft (G)
4.	Macdonald (G)	10.	Taylor (G)
5.	Pollock (W. I.)	11.	Trapp (W. I.)
6.	Puebla M. (?)	12.	Wagner (G)

13. Waldin (W. I.)

The above list is not in any order of preference but is only alphabetical. It is possible that in the opinion of individual growers some of the varieties included above may deserve to be replaced by others, which, in their estimation, have better qualities. The present list, however, is based on a consensus of opinion and not on individual likes and dislikes. It must be added that some of the recent seedlings from the Collins, Winslow, and Taylor are considered to hold great promise, though it is obviously too early yet to speak of them as standard varieties. It will be admitted that none of the pure seedlings of the West Indian, Guatemalan or Mexican avocado can by themselves produce the best varieties commercially, since whatever

defects exist are the defects of the group or "race" and are
therefore more or less reproduced in the pure line. If cross-
breeding is thus indispensable, it must be selective and not
left to chance. The varieties so far tested out in Florida may
therefore furnish at least a starting point for combining the
qualities desired.

The Trapp, Valdin, and Pollock furnish three of the
best West Indians, each remarkable for some particular quality:
the Trapp for its over-abundant crops, the Waldin for lateness
of maturity, (that is, in its class); and the Pollock for its
large size.

The Guatemalan group furnishes varieties, which, in
addition to their general hardiness and good shipping qualities,
include the Collins, Macdonald and Sharpless, noted for their
late season of maturing; and the Linda noted for its size.

Of the Mexican group, the varieties that particularly
merit consideration under Florida conditions are the Puebla and
the Fuerte, both of which are supposed to be Mexican hybrids;
the Fuerte, at least, is certainly of mixed parentage.

(Vide: Popenoe, F. W., Manual of Tropical and Subtropical
Fruits, 1920 p. 78, and California Avocado Association, Circular
No. 1, Oct. 25, 1917; also Annual Report on the California
Avocado Association, 1919, note on page 74).

The Knight, which is a Guatemalan and also considered
a very desirable variety, deserves a place in the above list.

Column legend (vertical headers, left to right):

- Hardy(H), medium(M) or tender(T)
- Season: early(E), Medium(M) or Late(L)
- Quality — Flavor: Good(G), Medium(M) or Poor(P)
- Flesh; fibrous(F), Not fibrous(N)
- Bearing quality; Good(Prolific)(G), Medium(M) or Poor(P)
- Size of fruit: Large(L), Medium(M) or small(S)
- Skin of fruit: Smooth(Sm), Medium(M) or Rough(R)
- Thick(T), Medium(M) or thin(t)
- Shape of fruit: Oblong, obovate(O), Round(R) or Pyriform(P)
- Seed — Tight in cavity(T), Medium(M) or loose(L)
- Large(L), Medium(M) or small(S)
- Ripening of fruit, even(E) or not even(N)
- Tree vigorous grower(G), Medium(M) or poor grower(P)
- Disease susceptible(S) or not susceptible(N)
- Bearing of Fruit: Early or young(E), Medium(M) or late(L)
- Regular(R), Medium(M) or not regular(N)
- Color of fruit, Green(G), purple(P)
- Remarks

Skin thick	Shape	Seed tight	Seed size	Ripening	Tree vigor	Disease	Bearing	Regular	Color	Remarks
T	O-F	T	M	N	G	N	E	R	G	Good late fruit, but small
M	O	T	M	N	G	S	E	R	G	Prolific but small and sometimes ripens unevenly
T	l	T	M	E	G	N	M	M	G	Good fruit. Tree disease resistant
T	O	T	M	E	G	N	E	M	P	A large fruit of more than medium quality
T	R	T	M	E	M	N	E	R	F	Excellent late fruit but rather small
M	O-P	L(?)	L	R	G	N	E	M	G	Large but early W. I.
t	O	T	S	B	G	N	E	R	P	
T	O-F	T	S	R	G	N	L	M	P	
T	P	T	M	E	G	N	M	R(?)	P	Good late fruit
T	O	T	M	B	G	N	M	M	G	Important commercial variety in Florida
T	P	T	M	E	M	N	E	R	G	Average commercial variety

It is omitted since it was not available for pollination at
the time.

5.   Below is an analysis of the characters of the different varieties
herein considered:

See tabular statement A.

6.   The qualities that we seek to counterbalance are these:

FRUIT	CROPS	SKIN
Early vs. late	Alternate vs. regular	Rough vs. smooth
Small vs. large	Shy vs. prolific	Tender vs. hard
Med. rich vs. super rich	Tree, slow growing vs. vigorous	

HARDINESS:

West Indian vs. Guatemalan for southern Florida

Mexican vs. Guatemalan for northern Florida and
for California

7. Some combinations attempted and the reasons for them:

1. Collins x Pollock

2. Fuerte x Linda
              x Taft
              x Queen
              x Knight

3. Knight x Trapp
           x Waldin
           x Fuerte
           x Puebla

4. Linda x Trapp
           x Waldin
           x Puebla
           x Fuerte

5. Macdonald x Pollock

6. Pollock x Collins
            x Macdonald
            x Wagner

7. Puebla x Taylor
        x Taft
        x Linda
        x Queen
        x Knight

8. Queen x Trapp
        x Waldin
        x Puebla
        x Fuerte

9. Sharpless x Trapp
           x Waldin

10. Taft x Trapp
        x Waldin
        x Puebla
        x Fuerte
        x Pollock

11. Taylor x Puebla
          x Waldin

12. Trapp x Linda
         x Queen
         x Knight
         x Taft

13. Wagner x Pollock

14. Waldin x Linda
          x Knight
          x Taft
          x Queen

Explanatory:

It may be well to state at the very outset that the object in making these combinations is twofold: It is obvious that these crosses cannot be expected to produce all at once the ideal fruit. But in the result one of two things must happen: either, in this first generation, as is likely to happen, the hybrid may contain a combination of characters intermediate between the two parents, in which case we shall have combinations which bring us nearer the ideal and which will be satisfactory commercially at least for the time being; or, the product of these crosses may cause the parent types (which may themselves be heterozygous) to break up into new combinations, in which case we shall have ascertained more definitely the behaviour

of these varieties in cross-pollination and also obtained
fresh material with which to build up the ideal fruit by a
series of further crosses.

The Pollock and the Collins are both very desirable
in their respective classes, but yet,they have excellent con-
trasting qualities. The first, being a West Indian, will tend
to be more hardy in union with the Guatemalan. The outstanding
defect of the Collins is that it is a trifle too small. It is,
however, a prolific variety. The Pollock, on the other hand,
is a fruit known for its size but it is not always a regular
bearer. Again the Pollock shares the defect common to the West
Indians in that its season is too early, while the Collins, if
anything, is a trifle too late. The Collins is considered ex-
ceptionally rich in flavor,while the Pollock would not be the
worse for having its flavor somewhat enriched. The Pollock,
partaking of the qualities of the harder skin of the Collins,
must prove a better shipper, while the shell-like skin of the
Collins must improve in appearance when moderated with the smoothness of the West Indian Pollock. It will thus be seen that the
Collins and the Pollock make an ideal combination. so far as
Florida growers are concerned. A hybrid that is fairly intermediate between the two parents would reasonably be expected to
bring us nearer the ideal.

The Macdonald stands almost precisely in the same
line as the Collins in relation to the Pollock.

The Taft, despite the slightly unfavorable reports
concerning it by growers too close to the sea, is by general
consent regarded as one of the best commercial Guatemalans in
Florida. The Waldin is regarded as its counterpart in the
West Indian group. These two varieties have many merits in
common and in the slight demerits they supplement each other.
The Taft is a fruit medium to late in maturing, while the
Waldin is a conspicuously late variety of the West Indian class,
which, however, is none too late for market requirements. The
Taft may welcome a slight increase in its size which the Waldin
can impart to it. The Waldin might be rendered hardier with
the Guatemalan blood in it, while the Taft would have much to
gain from the Waldin as a regular and prolific bearer. In
regard to the quality of the skin, the benefit is mutual; -
that of the Waldin being smooth but a somewhat delicate type,
while that of the Taft is rough but harder and ships better.

Among the other Guatemalan varieties of especial merit
are the Taylor, Queen, Wagner, Knight, Linda and Sharpless.
The performance records of the Knight and Sharpless are as yet
incomplete, but so far as is known at present under Florida
conditions, they promise a prominent place in commercial planting
for these varieties.

The Linda, in particular, is already noted for its
vigorous habit, while the fruit is conspicuous for its large
size. It is also a late fruit, but the bearing qualities are

not quite so uniform as one would wish them to be. The Trapp
is regarded as the complement of the Linda among the West
Indians, as it is notorious for its overprolific qualities,
irrespective of seasonal conditions. The early maturing quality
of the Trapp would be counterbalanced by the late habit of the
Linda and a blending of the somewhat rough skin of the Linda
with the smoothness of the Trapp would be a welcome feature.

The Wagner is considered an excellent Guatemalan,
its only defect being that it is a trifle too small. Crossed
with the Pollock, the result may bring us nearer the standard.

8. **Details of Pollination**

Before proceeding to describe the details of pollination
it will perhaps be well to outline the structure of the avocado
flower.

The following is a description of it by Wilson Popenoe in
the Cyc. of Horticulture 1916, Vol. V. Page 2556.

"The flowers of the avocado are shortly pedicellate in
broad compact panicles at the ends of the young branchlets about
3/8 inch across, greenish, the calyx lobes oblong lanceolate,
acute, slightly concave, finely pubescent; fertile stamens 9 in 3
series, each stamen of the inner series bearing just about its
base two oval, flattened, orange-colored glands; filaments slender,
finely hairy, the anthers oblong ovate, dehiscing by four valves
hinged distally, the two outer series dehiscing extrorsely, the
inner series with the two distal valves extrorse and the proximal
pair introrse; staminodes 3, flattened, orange-colored, ovary
ovate elliptic, the style slender, attenuate, finely pubescent."

The number of flowers in a panicle varies largely
with the variety. Approximately, about 150 to 500 flowers
are borne on the several panicles, emerging from each branchlet
and all these can be conveniently enclosed together in fairly
large-sized manilla bags (size 8). The length of time each
blossom remains open depends on the variety and is about
six to twelve hours. The bagging is done the previous day, about
18 to 20 hours previous to pollination after all the flowers
that should open that day have fully opened and have been
removed. When the bag is opened for pollination the next day,
usually about 10 to 30 fresh flowers will be found open. The
stigma is receptive before the pollen of the same flower is
released. The flowers are operated on as soon as possible
after anthesis and before the anthers have opened their valves
and shed their pollen. Since each flower has 9 anthers to be
removed and 30 flowers under a bag would entail the removal of
30 x 9, or 270 anthers, emasculation and disbudding the unopened
flowers necessitates considerable interval before replacing the
bag. It is safer, therefore, to pollinate first as soon as the
bag is removed and then proceed to emasculate.

A significant fact noticed in this investigation is
that the time of day at which the flowers open and close is
definite for each variety of the avocado, and that it is only
slightly modified by the weather conditions. Thus, in Florida,
in February or March on mornings which are cooler than usual and
when the temperature is about 45° to 50°, or on a very cloudy

day, the flowers will open from a few minutes to about an hour
later. Relatively, however, the succession in which they open
and close and complete their own cycle is surprisingly uniform
for each variety.

Thus on a bright day in February or March when the
temperature is around 70°, the Puebla blooms will open at 6 A. M;
those of the Taft at 8 A. M., the Waldin at 9 A. M., the Macdonald
at 1 P. M., the Linda at 3, the Pollock at 4, and the Trapp at
5 P. M.. Likewise, they shed their pollen at almost a definite
time of day: the Linda at 7, the Trapp at 8, and the Fuerte at
9 in the morning; the Puebla at 1, the Taft at 2, and the
Macdonald at 4 P. M. The Fuerte, Linda, Pollock, Queen and Trapp,
which open in the afternoon, close in the evening (as do also
the others) but without shedding their pollen. These shed their
pollen the next morning at a definite hour, and with the exception
of the Fuerte, close during the day at a definite time anterior
to the opening of fresh buds. The Linda and the Macdonald begin
to close at 11 A. M., the Pollock and the Queen close at 12 noon,
the Taft at 4 P. M., and the Taylor at 5 P. M. These observations
were recorded from day to day for over a month.

With such a rhythm and regularity in the opening and
closing of the flowers of each variety, it is not difficult with
the aid of a chart showing the progress of each, to ascertain the
definite time of day when fresh pollen can be secured for pollination,
and also the time in each case when pollination should be effected.

It is also fortunate that emasculation can be accomplished without
disturbing the buds before they have opened of themselves. When
the stigma is fully receptive, the stamens spread out to a
position at right angles to the style.and at this stage the
anthers can be removed easily with the tip of small,long handled
surgeon's scissors or with slender tipped embroidery scissors,
without cutting away the perianth lobes and without hurting the
flower unduly. At a later stage the perianth lobes become
somewhat recurved before closing up, more distinctly so in the
Fuerte. When the stigma is past receptivity it shows a brown-
ing of the tip. This fact, as also the open anther valves, can
easily be detected with a hand lens,and later when the eye is
trained to the shape of the anthers and of the stigma at different
stages, their condition can be determined even with the naked eye.
Since the stigma in the majority of cases observed shows a browned
and withered appearance before the anthers of the same flower
have opened their valves, the chances of self-pollination seem
remote.

There are cases, however, in which the three stamens
of the innermost series close up on the stigma in advance of
the release of pollen, and protect the stigma from withering up
by too prolonged exposure; in these cases it is possible that
the stigma is self-pollinated. This is rendered more probable
by the fact that the two proximal valves of this particular inner
series of anthers open introrsely. An alternative to this
conclusion is that the closing up of the inner series of stamens

Season of Bloom	Name of Variety	6 A.M.	7 A.M.	8 A.M.	9 A.M.						
From middle of Feb, into March	Fuerte	Fls. open : pollen : not shed	Fls. open : pollen : not shed	Fls. open : pollen : not shed	Fls. open : pollen : shed						
From middle of March	Linda	Fls. open : pollen : shed	Fls. open : pollen : shed	Fls. open : pollen : shed	Fls. open : pollen : shed						
From middle of March	Macdonald	Fls. open : pollen : not shed	Fls. open : pollen : not shed	Fls. open : pollen : shedding	Fls. open : pollen : shed						
From middle of Feb, into March	Pollock	Fls. open : pollen : not shed	Fls. open : pollen : not shed	Fls. open : pollen : shed	Fls. open : pollen : shed						
From middle of Feb, into March	Puebla	Buds : just : opening	Buds : just : opening	Fls. open : pollen : not shed	Fls. open : pollen : not shed	Fls. open : pollen : not shed	Fls. open : pollen : not shed	Fls. open : pollen : not shed	Fls. open : pollen : shedding	Fls. open : pollen : shed	Fls. open : pollen : shed
From March	Queen	Fls. open : pollen : shed	Fls. open : pollen : shed	Fls. open : pollen : shed	Fls. open : pollen : shed	Pollen : shed,fls. : closing	Pollen : shed,fls. : closing	Pollen shed : fls. nearly : closed	Fls. closed : fresh buds : opening	Fresh flow-: ers open, : pollen not : shed	
From March	Taft	Fls. not : open	Fls. not : open	Buds : just : opening	Fls. open : pollen : not shed	Fls. open : pollen : not shed	Fls. open : pollen : not shed	Fls. open : pollen : not shed	Fls. open : pollen : shedding	Fls. open : pollen : shedding	Fls. open : pollen : shed
From March	Taylor	Buds : just : opening	Buds : just : opening	Fls. open : pollen : not shed	Fls. open : pollen : not shed	Fls. open : pollen : not shed	Fls. open : pollen : not shed	Fls. open : pollen : not shed	Fls. open : pollen : shedding	Fls. open : pollen : shed	Fls. open : pollen : shed
From middle of Feb, into March	Trapp	Fls. open : pollen : not shed	Fls. open : pollen : not shed	Fls. open : pollen : shedding	Fls. open : pollen : shed	Fls. open : pollen : shed	Fls. open : pollen : shed	Pollen : shed,fls. : closing	Pollen shed : fls. nearly : closed	Fls. : closed	Fls. : closed
From March	Wagner	Fls. not : open	Fls. not : open	Buds : just : opening	Fls. open : pollen : not shed	Fls. open : pollen : not shed	Fls. open : pollen : not shed	Fls. open : pollen : not shed	Fls. open : pollen just : shedding	Fls. open : pollen : shed	Fls. open : pollen : shed
From middle of Feb, into March	Waldin	Fls. not : open	Fls. not : open	Fls. not : open	Buds : just : opening	Fls. open : pollen : not shed	Fls. open : pollen : not shed	Fls. open : pollen : shedding	Fls. open : pollen : shedding	Fls. open : pollen : shed	Fls. open : pollen : shed
From middle of March	Sharpless	Fls. not : open	Fls. not : open	Fls. open : pollen : not shed	Fls. open : pollen : not shed	Fls. open : pollen : not shed	Fls. open : pollen : shedding	Fls. open : pollen : shed	Fls. open : pollen : shed	Fls. open : pollen : shed	

Remarks: It is by no means claimed that this chronological statement is true irrespective of climatic conditions. It is possible that in different local areas the hours of opening and closing of flowers may be different for the same variety. But from what has been observed, there can be little doubt that there is nevertheless the same rhythm and regularity in the succession of flowers and the same cycle of progress for each variety in relation to the others.

around the stigma is an evidence that it has been fertilized
and is preparatory to the closing up of the entire flower, just
as happens normally in the case of the others after they have
shed their pollen.

This point must be left over for a future investigation
since want of facilities prevented the preparation of histological
sections of these flowers at the time. In any event, this does
not vitiate the attempt to pollinate by artificial means. In
every case it is possible to handle the flowers for the operation
before the closing up of the inner series and before any of its
own pollen has been released.

9.   With the aid of the tabular statement appended hereto, (Vide
statement B) it is possible, in the case of the varieties herein
considered, to determine the exact time, correct to about 15 minutes,
when the pollen of a desired variety should be taken in order to
obtain it as fresh as possible, and when pollination should be
effected in each case.

Thus, the pollen of the Waldin should be secured at
2 P. M., of the Taft and Taylor at 3 P. M., of the Macdonald,
Queen, Fuerte, Pollock, and Trapp at 9 A. M. Likewise, the
Wagner should be pollinated at 8 A. M., the Puebla at 9 A. M.,
the Queen at 2 P. M., the Fuerte, Macdonald and Pollock at
3 P. M., the Linda at 4, and the Trapp at 5 P. M.

The time indicated in the statement as "pollen shedding"
is naturally also the time at which it is secured for pollination
before it is blown off by the wind, but it will be found in many

cases that the time of day at which a particular variety should
be pollinated synchronises with the time at which pollen of the
desired variety is just shed and is directly obtainable from the
tree itself without having to store it in advance. In other
cases, however, it has to be secured either the same day, a few
hours previously, or on the previous evening. Thus, in pollinating the Wagner with either the Trapp or Macdonald, the pollen
can be obtained directly from the open blossoms of the latter,
but the pollen of the Valdin for the Wagner has to be secured
the previous evening at 2 P. M. For pollinating the Trapp at
5 P. M., the Queen pollen should be secured at 9 the same morning,
while that of the Taft is available just two hours previously
(i. e., at 3 P. M.).

For pollinating the Macdonald, the pollen of the Pollock
should be obtained between 8 and 9 the same morning, and in the
case of the Pollock, the pollen of the Macdonald should also be
secured at the same hour.

10.   In about three days after pollination, after all danger of the
emasculated flowers receiving pollen from the outside is long
past, it is well to puncture little holes in the bags in order to
admit light and air into them. In the present experiments, it
was found that three weeks after pollination in about 50 per cent
of the bags, from among the ten to thirty flowers pollinated under
each, one or two at least had set and in some as many as six to
eight.

11. **Supplementary**

As a by-product of this investigation (so to speak)
it is obvious that certain varieties interplanted with certain
other varieties ought to give better chances for the setting
of fruit.

Since the chances of self-pollination are remote,
in an orchard where interplanting is not done, the chances of
pollination and the setting of fruit under a natural process
are limited by the following adverse factors:

I. The pollen available for fertilization on any day is
the pollen of the flowers of the previous set which
may have survived adverse weather conditions in the open
field until the next day.

II. Most of the pollen that may thus survive is blown off
by the wind before a fresh succession of flowers have
had time to open.

III. The major portion of what is left over is enclosed
within the flowers which close up before fresh buds open.

The only exception to this, so far as could be observed,
is the Fuerte whose flowers remain open with the pollen exposed
while a fresh series of buds are opening. It is thus the only
variety in which the chances of pollination are not decreased
for want of interplanting.

Where interplanting is practised, it seems natural to
conclude that if other factors do not intervene, the chances

of the setting of fruit are greatly enhanced by planting together
varieties whose flowers when they open can be pollinated by
varieties whose flowers opened one or two hours previously and
whose pollen has been just released. Thus, we know (vide: statement B.) that the Linda opens its buds between 3 and 4 P. M.,
while the Waldin which opened its flowers at 9 A. M., has its
pollen released at that hour, and is available for the Linda
until the flowers close in the evening. Likewise, the Waldin,
when its buds are open at 9 A. M., has the pollen of the Linda
available for it since the latter is shed early in the morning
and its flowers remain open till 12 noon. In their season of
bloom, although the Waldin normally comes into bloom about two
weeks earlier, the latter part of its flowering season coincides
with that of the Linda.

We, therefore, know that so far as mutual benefit
for the setting of fruit is concerned, the Linda and Waldin
make an excellent combination for interplanting and from the
standpoint of the grower, also, these two varieties give him a
continuity of season in the maturing of fruit.

Likewise, a reference to statement B. will show that
the Waldin and Macdonald are a good combination. Other combinations for interplanting which are mutually beneficial are:

1. Fuerte & Macdonald
   " & Puebla
   " & Sharpless
   " & Taft
   " & Taylor
   " & Waldin

2. Linda & Puebla
   " & Sharpless
   " & Taft
   " & Taylor
   " & Wagner

3. Macdonald & Fuerte
   " & Taft
   " & Taylor
   " & Puebla
   " & Wagner
   " & Sharpless

4. Pollock & Puebla
   " & Taft
   " & Taylor
   " & Wagner
   " & Waldin
   " & Sharpless

5. Puebla & Fuerte
   " & Linda
   " & Macdonald
   " & Pollock
   " & Queen
   " & Trapp

6. Queen & Puebla
   " & Taft
   " & Taylor
   " & Wagner
   " & Waldin
   " & Sharpless

7. Sharpless & Fuerte
   " & Linda
   " & Macdonald
   " & Pollock
   " & Queen
   " & Trapp

8. Taft & Fuerte
   " & Linda
   " & Macdonald
   " & Pollock
   " & Trapp

9. Taylor & Fuerte
   " & Linda
   " & Macdonald
   " & Pollock
   " & Queen
   " & Trapp

10. Trapp & Puebla
    " & Taft
    " & Taylor
    " & Waldin
    " & Sharpless

11. Wagner & Linda
    " & Macdonald
    " & Pollock
    " & Queen

12. Waldin & Fuerte
    " & Pollock
    " & Queen
    " & Trapp

Among the varieties considered in this paper, combinations other than those mentioned above are either of no value for interplanting or the benefit is not mutual, but one-sided.

12. Summary

The avocado, because of its high food value, is
a fruit of immense possibilities and already a fruit of
commercial importance.

Because of the inadequate supply at present, the
avocado is bringing incredible prices in the northern markets.

Avocado orcharding in Florida already bids fair to
rival the citrus industry.

From a strictly commercial standpoint the industry
needs to be stabilized by confining production to one or two
standard varieties. The standard varieties which should meet
the requirements of the grower as well as the market have to be
evolved by cross-pollination.

The avocado consists of three distinct groups or "races"
called the West Indian, the Guatemalan and the Mexican. Of
these, the standard fruit of the future should be a cross between
the West Indian and the Guatemalan for southern Florida, and
the Guatemalan and Mexican for northern Florida and California.

The varieties now in cultivation contain among them
all the best qualities of the ideal commercial type of fruit,
and what is now needed is a combination of these qualities in
desirable hybrids by a succession of cross-pollinations.

An analysis of the characters of the leading varieties
now in the trade indicates what varieties should be cross-
pollinated in order that a start be made in this direction.

A study of the flowers of the different varieties during the several stages of anthesis reveals to us facts of far-reaching importance:

These are:

1. That the time of day at which the flowers open, shed their pollen and close, is distinct and definite for each variety, but is the same for all trees of the same variety from day to day.

2. That a tabular statement showing the different stages of the opening, maturity and closing up of the flowers for each variety gives us the exact time at which pollination should be carried on in each case.

3. This tabular statement also tells us that certain varieties should be interplanted with certain others in order to increase the chances of the setting of fruit.

The investigation thus suggests the means of producing a better avocado, nearer the ideal fruit of commerce, and also the means of producing heavier and more regular crops by a more careful system of interplanting different varieties.

* * * * * * * * * *

Massachusetts Agricultural College

Amherst, April 25, 1922

B.A.,F.R.H.S.

Appendix A.

"The following tabular statement shows conclusively how much richer than other fruits is the avocado in protein, mineral matter and total solids - in other words, contains far less water.

		Avocado		Other fruits
Water........	60-80	Average	70	82.50
Protein......	1.3-4.6	"	2.50	0.72
Ash..........	1.38-1.72	"	1.50	0.51

In this connection it is also of interest to compare the avocado with some other foods. The avocado shows on an average about 70 per cent of water; the potato, either raw or cooked, will vary from 75 - 78 per cent. The protein in the two foods is about the same with the advantage on the side of the fruit.

Raw cereals yield from 10 to 12 per cent of protein; when cooked, however, will average about 2½ per cent; rice, less than 2 per cent. The mineral matter of the avocado will average at least 1.5 per cent; the corresponding figure for the potato is 1.0, either raw or cooked; cooked rice, 0.15; cooked cereals, 0.5; meat 1.0; eggs 1.0. It is thus seen that the avocado when compared with cooked foods, which is the correct method of comparison, in view of the fact that we are discussing the material ready for consumption, is:

1. A richer source of mineral matter.

2. A richer source of protein than the potato,
   green vegetables or the cooked cereals.

The main constituent of the foods mentioned is the
carbohydrate, while the main constituent of the avocado is
the oil. The maximum percentage of carbohydrate in any of
the foods in question, ready for consumption, does not exceed
20 per cent, while the minimum percentage of oil noted for the
avocado is 9 with a maximum of 32. As the caloric value of
fat is 2 1/4 times that of the carbohydrates, it is obvious
that the caloric value of a unit weight of the edible portion
of the avocado is far greater than a corresponding weight of
any of the foods above discussed.

This places the avocado, as previously stated, in a
class by itself . . . . . "

Prof. Jaffa, M. B., Annual Report of the Cal.
Avocado Association, 1920-21, pp. 62-63.

Appendix B.

Comparative Hardiness of Varieties

*   *   *   *

"The following statements of temperature endurance
is based on a very large number of observations made in different
places and is as nearly correct as can be determined from the
data collected:

30°F. - Nothing injured so far as could be observed.

29°F. - No injury of account; only traces on most tender
         growth of West Indian and Guatemalan varieties.

28°F. - New foliage scorched on Guatemalan types; West
         Indian varieties showing considerable foliage
         damage.

27°F. - Mexican varieties, with new tips slightly scorched;
         Guatemalan, with almost all new foliage injured;
         West Indian, badly damaged.

25° to 26°F. - Mexican varieties, with new foliage injured, but
         some dormant trees uninjured; all Guatemalan
         sorts, with new foliage badly injured and some
         old foliage scorched.

24°F. - Some dormant Mexicans uninjured; Guatemalan
         varieties badly injured, small limbs frozen back.

21°F. - All Guatemalan types killed to bud; a few of the
   hardiest Mexicans, ... with young leaves
   only, injured.

It must be remembered that the above statements at
best can only be approximately correct; and much variation will
always be found, due to tree condition and environment."

Dr. Webber, H. J., Annual Reports of the California
Avocado Association, 1917, pp. 50-51.

Bibliography

Church, C. G. and Chace, E. M. On the maturity of avocados.
    Cal. Avocado Association, Annual Report 1921, pp. 45-59.

Condit, I. J. History of the avocado and its varieties in
    California. Cal. State. Com. Hort. B. V. 6: 1-21 Jan. 1917

Condit, I. J. Avocado, aristocrat of fruits. Amer. Fruit Grower
    39: 40, Sept. 1919

Elliott, J. M. Experimental work on certain avocados. Cal.
    Citrograph 5: 360 Sept. 1920

Harris, J. A. and Popenoe, W. Freezing point lowering of the sap
    of the horticultural varieties of Persea Americana. Journ.
    Agr. Research Vol. VII. No. 6, 1916, pp. 261-268

Higgins, J. E. The avocado in Hawaii. Hawaii Agr. Exp. Station
    Bul. No. 25

Hodgson, R. W. Avocado monstrosity. Journ. Heredity 8: 557-8,
    Dec. 1917

Jaffa, Prof. M. E., Nutritional studies of the avocado. Cal.
    Avocado Association, Annual Report 1921

LaForge, F. B. D - mannoketoheptose: a new sugar from the avocado.
    Franklin Institute, Philadelphia 184: 122, July 1917

Popenoe, P. O. Varieties of the avocado. Cal. Avocado Association,
    Annual Report 1915, pp. 44-68.

Popenoe, W. Manual of tropical and sub-tropical fruits. 1920.

-31-

Popenoe, W. Avocado in Guatemala. U. S. Ag. B. 743: 1-69, 1919

Popenoe, W. Avocados as food in Guatemala. Journ. Heredity 9:
    99-107, March 1918

------------ Avocado in Trinidad and Tobago. Agri. News 19: 46,
    Feb. 7, 1920

Rolfs, P. H. Avocado culture. Fla. Ag. Dept. Quar. B. V. 26: 43-4,
    Apr. 1916

Scott, L. B. Comparative merits of the California avocado varieties.
    Cal. Avocado Association, Annual Report 1917, pp. 57-62.

Shamel, A. D. Performance records of avocados based on citrus
    experiments. Cal. Citrograph 5: 68, Jan. 1920

Shamel, A. D. Avocado tree records. Cal. Cultivator 56: 235,
    Feb. 19, 1921

------------- Hardy avocado. Cal. Cultivator 56: 276, Feb. 26,
    1921

(See paragraph 8.)

The structure of the avocado flower.

An Avocado Orchard

Courtesy of Mr. George B. Cellon
Tropical Grove, Miami, Fla.

The Trapp Avocado. Third Crop.
138 Fruits on Five-Year Old Budded Tree

Courtesy of Mr. George B. Cellon
Tropical Grove, Miami, Fla.

The Pollock Avocado

(about 3/4 natural size)

Courtesy of Mr. George B. Cellon
Tropical Grove, Miami, Fla.

A Guatemalan Avocado

Compare the shell-like skin of this fruit

with that shown on next page.

Photo by Webber, Dr. H. J.
Cal. Avocado Association
Annual Report 1916.

The Dickey Avocado (Guatemalan)
Compare the leathery skin of this fruit with
those shown on the previous and the succeeding pages.

Photo by I. J. Condit
Cal. Avocado Association
Annual Report 1916.

The Ganter Avocado (Mexican)

Compare the smooth skin of this Mexican avocado
with that of the Guatemalans on pages 35 and 36.

Photo by I. J. Condit
Cal. Avocado Association
Annual Report 1916.

One of the several hundred budded trees selected
by the writer for cross-pollination.

CPSIA information can be obtained
at www.ICGtesting.com
Printed in the USA
BVHW091628121118
532888BV00015B/531/P